ACE THE DATA SCIENCE
INTERVIEW

UNVEIL THE SECRETS OF 100 REAL QUESTIONS
FROM THE WORLD'S LEADING TECH GIANTS

AI PUBLISHING

How to Contact Us

If you have any feedback or inquiries,
we invite you to reach out to us at
contact@aipublishing.io.

Your insights and opinions are incredibly valuable to us,
and we are eager to hear from you.
Your contributions will greatly aid us in enhancing
the quality and relevance of our publications.

To access Python Codes and other materials
used in this book, please visit the following link:
https://www.aipublishing.io/book-Interview-DS

(Note: An order number will be required for access.)

About the Publisher

AI Publishing Company is dedicated to providing a global learning platform for students, beginners, small businesses, startups, and managers venturing into the realms of data science and artificial intelligence. Our mission is to facilitate the acquisition of crucial AI and data science skills through interactive, coherent, and practical resources.

Our offerings span from introductory courses in programming languages and data science to advanced modules in machine learning, deep learning, computer vision, big data, and more, featuring languages like Python and R, along with other AI and data science software.

At the heart of AI Publishing's mission is empowering learners to address digital challenges proactively by maximizing AI and data science's transformative potential. Our online content and eBooks are crafted by experts to provide up-to-date, clear, and comprehensive insights into AI and data science, dispelling common myths and misunderstandings.

We also extend consultancy and corporate training services in AI and data science, helping businesses enhance their teams' efficiency and navigate complex workflows.

Stay at the forefront of AI and data science innovation with AI Publishing. Whether you're starting from scratch or seeking to deepen your expertise, we are your go-to resource for learning and mastering AI and data science. For any inquiries or further information, please reach us at contact@aipublishing.io.

Please contact us by email at
contact@aipublishing.io.

AI Publishing is Looking for Authors Like You

Interested in becoming an author for AI Publishing?
Please contact us at author@aipublishing.io.

We are working with developers and AI tech professionals just like you, to help them share their insights with the global AI and Data Science lovers. You can share all your knowledge about hot topics in AI and Data Science.

Table of Contents

Introduction

In the dynamic and ever-evolving world of technology, data science has emerged as a pivotal field, driving innovation and decision-making in some of the most influential companies globally. "Ace the Data Science Interview: Unveil the Secrets of 100 Real Questions from the World's Leading Tech Giants" is a meticulously crafted guide designed to navigate the challenging and rewarding path of landing a data science role in these prestigious organizations.

This book is not just a collection of questions; it's a comprehensive journey into the heart of data science interviews conducted by the tech industry's titans - the FAANG companies (Facebook, Amazon, Apple, Netflix, and Google). These companies, known for their rigorous hiring standards and cutting-edge work, require candidates to possess a unique blend of technical expertise, problem-solving skills, and creative thinking.

Through "Ace the Data Science Interview," you will gain exclusive insights into the types of questions posed by these tech giants, understand the intricacies behind them, and learn how to approach, solve, and effectively communicate your solutions. Whether it's statistical inference, machine learning, algorithm design, or data analysis, this book covers a vast spectrum of topics that a data science candidate needs to be proficient in.

We recognize that cracking a data science interview at a FAANG company is about more than just technical know-how. It's about understanding the mindset behind the questions, showcasing your analytical abilities, and demonstrating how you can add value to some of the most innovative teams in the tech industry. Each question in this book is accompanied by detailed explanations, strategies for solving, and tips on what interviewers are looking for.

"Ace the Data Science Interview" is designed for a wide range of readers - from those taking their first steps into the world of data science to seasoned professionals seeking to take their careers to new heights. This book is your ally, whether you are preparing for your first data science interview or looking to brush up on your skills.

Embark on this journey with us, as we guide you through the nuances of acing a data science interview at a FAANG company. With "Ace the Data Science Interview," unlock the door to a world of opportunities in the exciting field of data science at the world's leading tech giants. Your path to becoming a part of these groundbreaking teams starts here.

Additional Resources to Accompany the Book

For comprehensive interview preparation and personalized coaching, we invite you to visit aisciences.io or ia email at contact@aisciences.io. Our platform is dedicated to equipping you with the tools, knowledge, and confidence needed to excel in your data science interviews. At aisciences.io, we understand the nuances of the interview process at top tech companies and offer tailored guidance to help you navigate these challenges successfully.

Our team of experienced professionals will work with you to sharpen your skills, refine your approach, and ensure you're fully prepared for every aspect of your interview. From deep dives into technical concepts to mastering the art of communication, aisciences.io is your partner in achieving interview success.

Preface

Welcome to "Ace the Data Science Interview: Unveil the Secrets of 100 Real Questions from the World's Leading Tech Giants," a comprehensive guide designed to empower aspiring data scientists in their journey towards landing a coveted position at one of the FAANG companies - Facebook, Amazon, Apple, Netflix, and Google.

The journey to a career in data science, particularly at a leading tech giant, is often shrouded in mystery and challenge. These companies, known for their innovative approach and highly competitive environment, seek the brightest minds in data science. As a result, their interview processes are rigorous, multifaceted, and sometimes daunting.

This book is born from the collective wisdom of industry experts, interview coaches, and successful candidates from these tech giants. It is meticulously crafted to demystify the interview process and provide you with an insider's view of what it takes to succeed. We have collated 100 real interview questions, ranging from statistics and programming to machine learning and problem-solving, ensuring a holistic preparation.

Our aim is not merely to present questions and answers; instead, we strive to offer a deeper understanding of the underlying concepts and thought processes that are essential in answering these questions effectively. Each question is accompanied by detailed explanations, practical tips, and insights into what interviewers are truly seeking.

In addition to technical prowess, this book also emphasizes the importance of soft skills - communication, problem-solving, and teamwork - which are equally crucial in securing and excelling in a data science role at a FAANG company. We provide guidance on how to articulate your thoughts, collaborate in team settings, and approach problem-solving tasks, ensuring you are well-rounded and interview-ready.

"Ace the Data Science Interview" is more than just a preparation guide; it is a companion in your journey towards achieving your dream job. Whether you are a budding data scientist or an experienced professional looking to take your career to new heights, this book is a valuable resource to help you navigate the complexities of the FAANG interview process.

As you turn these pages, we invite you to dive deep into the world of data science interviews, armed with knowledge, confidence, and a keen understanding of what it takes to succeed. Your journey to becoming a part of the world's most innovative tech teams begins here.

Wishing you the best in your endeavors,

AI Publishing

Book Approach

"Ace the Data Science Interview" adopts a holistic and comprehensive approach to preparing for FAANG data science interviews. Our method is twofold:

Firstly, we delve deep into the technical aspects. Each of the 100 real interview questions is presented with a clear, detailed explanation, not just of the answer, but of the underlying concepts. This approach ensures you grasp the fundamental principles, allowing you to adapt and apply this knowledge to a variety of scenarios.

Secondly, we focus on the strategic aspects of the interview process. Understanding the mindset of FAANG interviewers and the rationale behind each question type is crucial. We provide insights into what the interviewers are looking for in your responses, how they assess your problem-solving skills, and how you can demonstrate your capabilities most effectively.

Throughout the book, we emphasize the importance of a balanced skill set - combining technical knowledge with soft skills like communication, teamwork, and analytical thinking.

§ Who Is This Book For?

This book is designed for a wide range of audiences aspiring to break into the field of data science, particularly those aiming for roles in FAANG companies. It is ideal for:

- **Aspiring Data Scientists**: Beginners who are just stepping into the world of data science and looking to build a strong foundation.

- **Experienced Data Professionals**: Those who have been in the field and wish to advance their careers by joining top tech companies.

- **Graduates and Post-Graduates**: Students from data science, computer science, or related fields who are about to enter the job market.

- **Career Switchers**: Professionals from different backgrounds moving into data science roles, seeking to understand what top companies require.

- **Interview Preparers**: Anyone who is actively preparing for data science interviews and wants to refine their approach and techniques.

§ How to Use This Book?

To maximize the benefits of "Ace the Data Science Interview," we recommend the following approach:

- **Sequential Reading**: Start from the beginning and progress through the chapters. This helps build your understanding progressively as concepts and difficulty levels evolve.

- **Active Practice**: Don't just read the questions and answers; try solving each question on your own first.

Then, compare your approach with the provided solutions.

- **Note-Taking and Highlighting**: Mark important points, and take notes on concepts that need further review or practice.

- **Regular Review**: Revisit chapters and questions periodically to reinforce your understanding.

- **Mock Interviews**: Use the questions in this book to simulate real interview scenarios. Practice articulating your answers aloud, as if in an actual interview.

- **Leverage the Additional Resources**: Make use of the links, references, and further reading suggestions provided to deepen your learning.

- **Feedback and Iterative Learning**: Reflect on areas where you struggle and seek additional resources or mentorship to improve in those areas.

By following these guidelines, you can use this book not just as a study guide, but as a tool for comprehensive skill development in preparation for a successful career in data science.

1

Crafting the Perfect Data Science Resume for FAANG

The Importance of a Resume in Landing a Data Science Job at FAANG

In the quest to join a FAANG company as a data scientist, your resume is your first impression, a personal billboard that can either open the door to opportunity or close it. These tech giants, renowned for their rigorous selection processes, receive thousands of applications. A well-crafted resume does more than list your experiences; it highlights your journey, skills, and potential in a way that resonates with what these top-tier companies are looking for.

Principles for Crafting an Appealing Data Science Resume

1. **Clarity and Conciseness**: Your resume should be clear, concise, and to the point. Avoid unnecessary jargon and focus on what matters. Limit it to one or two pages.

2. **Tailor Your Resume**: Customize your resume for the role you are applying for. Emphasize the skills and experiences that align with the job description.

3. **Quantifiable Achievements**: Wherever possible, use numbers and data to demonstrate your impact. For instance, "Optimized algorithm performance by 30%."

4. **Relevant Technical Skills**: Clearly outline your technical toolkit – programming languages, data analysis software, machine learning techniques, etc.

5. **Education and Certifications**: Include your formal education, relevant coursework, and any additional certifications that enhance your suitability for the role.

6. **Project Experience**: Showcase key projects that align with the role. Explain the problem, your approach, the tools used, and the outcome.

7. **Professional Experience**: Highlight relevant job experiences, internships, or research positions. Focus on roles that showcase your data science capabilities.

Resume Hacks for Data Scientists

- **Use Industry Keywords**: Incorporate keywords from the data science field and the job description. This is especially important for passing automated resume screenings.

- **GitHub and Portfolio Links**: Include links to your GitHub repository or portfolio to provide a practical demonstration of your skills.

- **Leadership and Teamwork**: Demonstrate your ability to lead projects or work effectively in a team. Data scientists in FAANG companies often work collaboratively.

- **Continuous Learning**: Mention any ongoing learning or upskilling, like online courses or workshops. This shows your commitment to staying updated in the field.

- **Problem-Solving Focus**: Frame your experiences in terms of problems solved or insights gained. This highlights your analytical and problem-solving skills.

- **Visual Appeal**: Use a clean, professional layout. Tools like LaTeX or advanced word processors can help make your resume visually appealing.

- **Proofread and Feedback**: Ensure your resume is free from errors. Get feedback from mentors or peers, especially those with industry experience.

Remember, your resume is a strategic tool in your job hunt – it's your narrative crafted to capture the attention of some of the most innovative companies in the world. By adhering to these principles and utilizing these hacks, you can elevate your resume from a mere formality to a compelling story of your professional journey in data science.

2

Statistics and Probability

1. What is a p-value and what does it signify about the statistical data?

A p-value is a measure used in statistical hypothesis testing to determine the significance of the observed data. Specifically, it represents the probability of obtaining results at least as extreme as the observed ones, assuming that the null hypothesis (a default position that there is no effect or no difference) is true. A low p-value (typically ≤ 0.05) indicates that the observed data are unlikely under the null hypothesis and thus lead to its rejection, suggesting the alternative hypothesis may be true. It's important to note that a p-value does not measure the probability that either hypothesis is true.

2. Explain Bayes' Theorem.

Bayes' Theorem is a fundamental concept in probability theory. It describes the probability of an event, based on prior knowledge of conditions that might be related to the event. The theorem is mathematically stated as $P(A|B) = [P(B|A) * P(A)] / P(B)$, where:

- $P(A|B)$ is the probability of event A occurring given that B is true.

- P(B|A) is the probability of event B given that A is true.

- P(A) and P(B) are the probabilities of observing A and B independently of each other. Bayes' Theorem is widely used in various fields, including data science, for updating the probability as more evidence becomes available.

3. What is the Central Limit Theorem and why is it important?

The Central Limit Theorem (CLT) states that the distribution of sample means approximates a normal distribution (bell-shaped curve), regardless of the shape of the population distribution, provided the sample size is sufficiently large and the samples are independent. This theorem is crucial because it allows for various statistical methods and conclusions to be applied, even when the population distribution is unknown, as long as the sample size is large enough (usually n > 30). The CLT is the foundation for many inferential statistics methods, including hypothesis testing and confidence intervals.

4. Describe the difference between Type I and Type II errors.

In statistical hypothesis testing, a Type I error occurs when the null hypothesis is true but is incorrectly rejected. It's also known as a "false positive." A Type II error, on the other hand, occurs when the null hypothesis is false but erroneously fails to be rejected. This is known as a "false negative." The significance level (alpha) of a test is the probability of making a Type I error, while the power of a test is related to the probability of not making a Type II error.

5. How do you interpret a ROC curve?

A ROC (Receiver Operating Characteristic) curve is a graphical representation used to evaluate the performance of a binary classifier system. It plots two parameters: True Positive Rate (TPR, or sensitivity) against False Positive Rate (FPR, or 1 - specificity). The curve shows the trade-off between sensitivity and specificity (any increase in sensitivity will be accompanied by a decrease in specificity). The closer the curve follows the left-hand border and then the top border of the ROC space, the more accurate the test. The area under the ROC curve (AUC) is a measure of the test's overall performance.

6. Explain the concept of statistical power.

Statistical power is the probability that a statistical test will correctly reject a false null hypothesis (detect a true effect). It is directly influenced by several factors, including the sample size, the effect size, the significance level, and the variability in the data. High statistical power means a greater probability of detecting an effect if there is one, thus reducing the likelihood of a Type II error. Power analysis is often used to determine the minimum sample size needed to detect an effect of a given size with a certain degree of confidence.

7. What is the difference between correlation and causation?

Correlation and causation are distinct concepts. Correlation implies a statistical relationship between two variables; when one changes, the other tends to change in a predictable way. However, correlation does not imply causation; that is, it does not prove that changes in one variable cause changes in another. Causation indicates that a change in one variable is responsible for the change in the other. Establishing causation usually requires controlled experimentation and cannot be concluded from correlation alone.

8. How would you explain the Law of Large Numbers to someone who knows little about statistics?

The Law of Large Numbers is a principle that describes the result of performing the same experiment a large number of times. According to this law, the average of the results from a large number of trials will converge to the expected value as the number of trials increases. In simpler terms, it means that the more times you repeat an experiment, the closer the experimental results will be to the expected outcomes. For example, while flipping a coin a few times might not always result in a 50-50 split between heads and tails, flipping it many times will typically yield a result close to this expected distribution.

9. What is a confidence interval and how do you interpret it?

A confidence interval is a range of values, derived from the sample statistics, that is believed to contain the population parameter with a certain level of confidence. It is expressed as an interval with an upper and lower boundary and a confidence level (commonly 95%). For example, a 95% confidence interval for a population mean indicates that if the same population is sampled multiple times and intervals are calculated, approximately 95% of those intervals will contain the true population mean. It's a way to express uncertainty in estimation.

10. Describe a real-world scenario where a Poisson distribution is applicable.

A Poisson distribution is often used in scenarios where we are counting the number of times an event happens within a fixed interval of time or space. A real-world example would be counting the number of cars passing through a toll booth on a highway in an hour. This scenario fits the criteria for a Poisson distribution: the number of cars passing through is counted over a constant time interval, the occurrence of one car passing does not affect the chances of another car passing, and the average number of cars passing per hour is consistent.

Programming and Data Manipulation

11. Write a SQL query to find the second highest salary from a table.

```
SQL Code
SELECT MAX(Salary)
FROM Employees
WHERE Salary NOT IN (SELECT MAX(Salary) FROM Employees);
```

This query first finds the highest salary, then excludes it and finds the maximum of the remaining salaries, which is the second highest.

12. How do you handle missing or corrupted data in a dataset?

Handling missing or corrupted data can be approached in several ways depending on the context and extent of the issue:

Deleting Rows or Columns: Remove rows with missing data if they are not significant, or drop columns with a high percentage of missing values.

Imputation: Replace missing values with a central tendency measure (mean, median, mode) or use more complex algorithms like k-NN, MICE, or Deep Learning.

Flagging and Filling: Create a new variable to indicate missingness and fill missing values with a number like 0, a central value, or use a predictive model.

For corrupted data, identify and rectify the errors, which could involve removing, correcting, or replacing corrupted entries.

13. Write a Python function to sort a list of numbers without using built-in sort functions.

```
Python Code
def bubble_sort(arr):
    n = len(arr)
    for i in range(n):
        for j in range(0, n-i-1):
            if arr[j] > arr[j+1]:
                arr[j], arr[j+1] = arr[j+1], arr[j]
    return arr
```

This function uses the bubble sort algorithm to sort a list of numbers.

14. Explain the difference between a list and a tuple in Python.

List: A list in Python is a dynamic array. It can be modified after its creation (mutable), allowing changes like adding, removing, or changing items. Lists are defined using square brackets [].

Tuple: A tuple is similar to a list but is immutable. Once a tuple is created, its elements cannot be changed, added, or removed. Tuples are defined using parentheses ().

15. How would you merge two dataframes in Python using Pandas?

```
Python Code
import pandas as pd
# Assuming two DataFrames df1 and df2
merged_df = pd.merge(df1, df2, on='common_column')
```

This code merges df1 and df2 on a common column. The merge function can be customized with parameters like how (type of merge - inner, outer, left, right), on (common column), etc.

16. Demonstrate how to implement a binary search algorithm.

```
Python Code
def binary_search(arr, x):
    low = 0
    high = len(arr) - 1
    mid = 0

    while low <= high:
        mid = (high + low) // 2
        if arr[mid] < x:
            low = mid + 1
        elif arr[mid] > x:
            high = mid - 1
        else:
            return mid
    return -1
```

This function performs a binary search for x in a sorted array arr.

17. What are decorators in Python, and how would you use them?

Decorators in Python are a design pattern that allows the user to add new functionality to an existing object without modifying its structure. Decorators are typically used to extend the behavior of functions and methods without permanently modifying them. They are defined with the @ symbol followed by the decorator function name.

```python
Python Code
def my_decorator(func):
    def wrapper():
        print("Something is happening before the function is called.")
        func()
        print("Something is happening after the function is called.")
    return wrapper

@my_decorator
def say_hello():
    print("Hello!")

say_hello()
```

18. Write a SQL query to join two tables.

```sql
SQL Code
SELECT *
FROM table1
INNER JOIN table2
ON table1.common_field = table2.common_field;
```

This query performs an inner join between table1 and table2 on a common field.

19. Demonstrate a simple map-reduce example in Python.

```python
from functools import reduce
# Map
data = [1, 2, 3, 4, 5]
mapped_data = map(lambda x: x * x, data)  # Square each
element
# Reduce
sum_of_squares = reduce(lambda x, y: x + y, mapped_data)
print(sum_of_squares)
```

This code first squares each element of the list (map step) and then sums them up (reduce step).

20. What are lambda functions in Python, and give an example of how you might use one.

Lambda functions in Python are small anonymous functions defined with the lambda keyword. They can have any number of arguments but only one expression.

```python
multiply = lambda x, y: x * y
print(multiply(5, 6))  # Output: 30
```

This lambda function takes two arguments x and y and returns their product.

21. Write a function in Python to find the mean of a list of numbers.

```python
def find_mean(numbers): return sum(numbers) / len(numbers) if
numbers else 0
mean_result = find_mean([1, 2, 3, 4, 5]) # Output: 3.0
```

22. How would you reverse a string in Python without using any built-in functions?

Python Code

```
def reverse_string(s): reversed_string = '' for char in s:
reversed_string = char + reversed_string return reversed_
string
```

Example Usage:

Python Code

```
reversed_string_result = reverse_string("hello") # Output:
'olleh'
```

23. Explain and implement a binary search algorithm in Python.

Python Code

```
def binary_search(arr, target): low, high = 0, len(arr) - 1
while low <= high: mid = (low + high) // 2 if arr[mid] <
target: low = mid + 1 elif arr[mid] > target: high = mid - 1
else: return mid return -1
```

Example Usage:

Python Code

```
# Assuming the array is sorted, for instance, arr = [1, 2,
3, 4, 5] # To search for a target value, say 4, call binary_
search(arr, 4)
```

24. Write a SQL query to find the top 3 highest salaries in an 'Employee' table.

SQL Code

```
SELECT Salary FROM Employee ORDER BY Salary DESC LIMIT 3;
```

25. How would you merge two sorted arrays into a single sorted array?

- **Explanation:** To merge two sorted arrays, we iterate through both arrays simultaneously, comparing their elements and appending the smaller one to the result array. We continue until we exhaust one array and then append the remaining elements of the other array.

- **Code:**

```python
Python Code
def merge_sorted_arrays(arr1, arr2): result = [] i, j = 0, 0
while i < len(arr1) and j < len(arr2): if arr1[i] < arr2[j]:
result.append(arr1[i]) i += 1 else: result.append(arr2[j]) j
+= 1 result.extend(arr1[i:]) result.extend(arr2[j:]) return
result
```

- **Test Output**: Merging **[1, 3, 5]** and **[2, 4, 6]** gives **[1, 2, 3, 4, 5, 6]**.

26. Implement a function to check if a given string is a palindrome.

- **Explanation**: A palindrome is a string that reads the same forward and backward. We can check this by comparing the string with its reverse.

- **Code**:

```python
Python Code
def is_palindrome(s): return s == s[::-1]
```

- **Test Output**: Checking **"radar"** returns **True**.

27. Describe how you would use regular expressions to find all the email addresses in a long text.

- **Explanation**: Regular expressions can be used to match patterns in text. An email address typically has a format that can be patterned, such as **username@domain. extension**.

- **Regular Expression Pattern**:

```
Python Code
import re pattern = r'\b[A-Za-z0-9._%+-]+@[A-Za-z0-9.-]+\.
[A-Z|a-z]{2,}\b'
```

- Use **re.findall(pattern, text)** to find all email addresses in **text**.

28. Write a Python function to implement the Fibonacci sequence.

```
Python Code
def fibonacci(n): a, b = 0, 1 sequence = [] for _ in range(n):
sequence.append(a) a, b = b, a + b return sequence
```

- **Test Output**: The first five numbers in the Fibonacci sequence are **[0, 1, 1, 2, 3]**.

29. How can you handle missing values in a dataset using Pandas?

Explanation: Pandas provides functions like **fillna()** to replace missing values with a specified value, and **dropna()** to remove rows with missing values.

```
Python Code
df_filled = df.fillna(0) # Example DataFrame df has missing
values
```

Test Output: Missing values in **df** are filled with **0** in **df_filled**.

30. Write a SQL query to count the number of orders each customer has in an 'Orders' table.

- **SQL Query**:

```
SQL Code
SELECT CustomerID, COUNT(OrderID) as NumberOfOrders FROM
Orders GROUP BY CustomerID;
```

Explanation: This SQL query counts the number of orders for each customer by grouping the records based on **CustomerID**.

31. Given an unsorted array of integers, write a function to find the length of the longest consecutive elements sequence.

Explanation: The function finds the longest streak of consecutive numbers in an unsorted array.

```
Python Code
def longest_consecutive(nums): if not nums: return 0 nums =
set(nums) longest_streak = 0 for num in nums: if num - 1 not
in nums: current_num = num current_streak = 1 while current_
num + 1 in nums: current_num += 1 current_streak += 1 longest_
streak = max(longest_streak, current_streak) return longest_
streak
```

Test Output: **4** for the array **[100, 4, 200, 1, 3, 2]**.

32. Implement a Python function to sort a list of strings by their length.

```
Python Code
def sort_strings_by_length(strings): return sorted(strings,
key=len)
```

- **Test Output**: **['kiwi', 'apple', 'banana']**.

33. Write a Python program to find the second largest number in an array.

Python Code

```python
def find_second_largest(numbers): first, second = float('-inf'),
float('-inf') for number in numbers: if number > first: first,
second = number, first elif first > number > second: second =
number return second
```

- **Test Output**: 4 for **[1, 3, 4, 5, 0, 2]**.

34. How would you extract the nth word in a string ?

Python Code

```python
def extract_nth_word(s, n): words = s.split() return
words[n-1] if 0 < n <= len(words) else None
```

- **Test Output**: **'a'** for "This is a test string" (3rd word).

35. Write a Python function that converts a decimal number to binary.

Python Code

```python
def decimal_to_binary(n): return bin(n).replace("0b", "")
```

- **Test Output**: **'1010'** for the decimal number **10**.

36. Describe a situation where you would use a left join in SQL.

Situation: Find all customers and their orders, including those who haven't ordered.

SQL Code

```sql
SELECT Customers.CustomerName, Orders.OrderID FROM Customers
LEFT JOIN Orders ON Customers.CustomerID = Orders.CustomerID;
```

37. Given two tables, write a SQL query to join them on a common column.

SQL Code

```
SELECT Employees.Name, Departments.DepartmentName FROM
Employees JOIN Departments ON Employees.DepartmentID =
Departments.DepartmentID;
```

38. Write a function in Python to count the occurrences of a specific character in a string.

Python Code

```
def count_character(s, char): return s.count(char)
```

- **Test Output**: **2** for counting 'o' in "hello world".

39. How would you implement a queue using two stacks in Python?

Python Code

```
class Queue: def __init__(self): self.stack1 = [] self.
stack2 = [] def enqueue(self, x): self.stack1.append(x) def
dequeue(self): if not self.stack2: while self.stack1: self.
stack2.append(self.stack1.pop()) return self.stack2.pop() if
self.stack2 else None
```

40. Write a Python program to check if two strings are anagrams of each other.

Python Code

```
def are_anagrams(str1, str2): return sorted(str1) ==
sorted(str2)
```

- **Test Output**: **True** for "listen" and "silent".

Machine Learning

 41. Explain the difference between supervised and unsupervised learning.

Supervised learning is a type of machine learning where the model is trained on labeled data. The training data includes input-output pairs, and the model learns to map inputs to outputs, making it suitable for tasks like regression and classification.

Unsupervised learning, on the other hand, deals with unlabeled data. The model learns patterns and structures from the input data without any explicit instructions on what to predict. It is used for clustering, association, and dimensionality reduction tasks.

42. What are the different types of recommendation systems?

There are primarily three types of recommendation systems:

- **Content-Based**: Recommends items similar to those a user has liked before, based on item features.

- **Collaborative Filtering**: Makes recommendations based on the preferences of similar users. This can be user-based or item-based.

- **Hybrid Systems**: Combine both content-based and collaborative filtering approaches to leverage the strengths of both.

43. How do you prevent overfitting in a model?

To prevent overfitting in a model:

- Use more data, if available, to provide a broader learning scope.
- Apply regularization techniques (like L1 or L2 regularization) to penalize complex models.
- Reduce model complexity by selecting a simpler model or reducing the number of features.
- Use techniques like cross-validation to evaluate model performance.
- Implement early stopping during training (common in neural networks).
- Use data augmentation techniques (especially in image processing tasks).

44. What is cross-validation, and why is it important?

Cross-validation is a technique used to evaluate the generalizability of a model by dividing the data into subsets and testing the model on each subset. The most common method is k-fold cross-validation, where data is split into k subsets, and the model is trained on k-1 subsets and tested on the remaining one, repeatedly. It is crucial because it provides a more robust measure of a model's performance compared to a single train-test split, especially when data is limited.

45. Explain the concept of ensemble learning.

Ensemble learning is a machine learning paradigm where multiple models (often called "weak learners") are trained to solve the same problem and combined to improve the overall performance. The main premise is that a group of weak learners can come together to form a strong learner, thereby increasing the accuracy and robustness of the model. Common ensemble methods include Bagging (Bootstrap Aggregating), Boosting, and Stacking.

46. Describe a real-world application of a neural network.

A real-world application of neural networks is in image recognition and classification. For instance, they are used in Facebook's facial recognition technology to identify and tag people in photos. Neural networks, especially deep learning models like convolutional neural networks (CNNs), have proven highly effective at interpreting and classifying visual data.

47. What are hyperparameters in machine learning models?

Hyperparameters are the configuration settings used to structure a machine learning model. They are external to the model and cannot be learned from the data. They are used to control the learning process and include settings like learning rate, number of hidden layers and neurons in neural networks, regularization parameters, and k in k-NN algorithms. Hyperparameter tuning is crucial for optimizing model performance.

48. How would you evaluate the accuracy of a logistic regression model?

The accuracy of a logistic regression model is typically evaluated using metrics like:

- **Accuracy**: The proportion of true results (both true positives and true negatives) in the total data set.
- **Precision and Recall**: Especially in imbalanced datasets.
- **ROC Curve and AUC Score**: Plotting the true positive rate against the false positive rate.
- **Confusion Matrix**: To visualize the performance of the algorithm.

49. What is the purpose of a cost function in machine learning?

A cost function, also known as a loss function, quantifies the error between predicted values and the actual values. The purpose of a cost function is to be minimized during the training process. This minimization drives the adjustments to the model's parameters, leading the model to make more accurate predictions.

50. Explain the difference between a random forest and a decision tree.

- **Decision Tree**: A decision tree is a single tree structure where each node represents a feature, each branch represents a decision rule, and each leaf node represents an outcome. It is simple to understand and visualize but often prone to overfitting.
- **Random Forest**: A random forest is an ensemble of decision trees, usually trained with the "bagging" method. It is more robust and accurate than a single

decision tree as it reduces the risk of overfitting by averaging multiple trees. It also handles both regression and classification tasks well.

Case Studies and
Problem Solving

51. How would you design an algorithm to predict customer churn?

To design an algorithm for predicting customer churn:

Data Collection: Gather customer data including usage, transaction history, support interactions, and demographics.

Feature Engineering: Identify key factors that influence churn, such as service usage patterns, customer satisfaction scores, and payment history.

Model Selection: Choose a suitable model like logistic regression, decision trees, or ensemble methods.

Training and Testing: Split the data into training and testing sets to train the model and validate its accuracy.

Evaluation: Use metrics like accuracy, precision, recall, and the ROC curve to evaluate the model's performance.

Iteration: Continuously refine the model by incorporating new data and feedback.

52. A/B Testing: How would you design and analyze an A/B test for a new product feature?

For A/B testing a new product feature:

- **Define Objective**: Clearly define what you are testing and the expected outcome (e.g., increased engagement).

- **Create Hypothesis**: Formulate a hypothesis on how the new feature will impact user behavior.

- **Segmentation**: Randomly divide your user base into two groups - one with the new feature (test group) and one without (control group).

- **Test Execution**: Implement the feature for the test group while monitoring both groups.

- **Data Collection**: Collect relevant data from both groups during the testing period.

- **Statistical Analysis**: Analyze the data using statistical methods to determine if there is a significant difference between the two groups.

- **Decision Making**: Decide whether to implement, modify, or abandon the new feature based on the results.

53. How would you approach identifying fraud in financial transactions?

To identify fraud in financial transactions:

- **Anomaly Detection**: Use statistical techniques to identify transactions that deviate significantly from the norm.

- **Machine Learning Models**: Implement models like supervised classification (e.g., logistic regression, SVM) to learn from past instances of fraud.

- **Pattern Recognition**: Look for patterns common in fraudulent transactions.

- **Real-Time Analysis**: Develop systems that can flag transactions in real time for further investigation.

- **User Behavior Analysis**: Monitor for unusual behavior in customer accounts.

- **Continual Learning**: Regularly update the model with new data to adapt to evolving fraud tactics.

54. Propose a strategy to optimize an e-commerce website's sales using data analysis.

To optimize sales for an e-commerce website:

- **Customer Segmentation**: Use clustering algorithms to segment customers for targeted marketing.

- **Personalized Recommendations**: Implement recommendation systems to suggest products based on past purchases and browsing behavior.

- **Price Optimization**: Analyze price sensitivity and competition to optimize pricing strategies.

- **Website Traffic Analysis**: Use web analytics to understand user behavior and improve the user journey.

- **Conversion Rate Optimization**: Analyze user interactions to identify and remove bottlenecks in the conversion funnel.

- **Sales Forecasting**: Use predictive analytics to forecast sales and manage inventory effectively.

55. Develop an approach to segment customers based on their buying behavior.

For customer segmentation based on buying behavior:

- **Data Collection**: Gather data on customer purchases, browsing history, and preferences.

- **RFM Analysis**: Implement Recency, Frequency, Monetary (RFM) analysis to understand customer purchasing patterns.

- **Clustering Algorithms**: Use algorithms like K-means to group customers into segments.

- **Behavioral Analysis**: Analyze behavioral data to identify patterns and preferences within each segment.

- **Segment Validation**: Validate the segments for consistency and stability over time.

- **Actionable Strategies**: Develop tailored marketing and sales strategies for each segment.

56. How would you use data to improve a company's social media strategy?

To improve a social media strategy using data:

- **Engagement Analysis**: Analyze likes, shares, comments, and reach to understand what content resonates with the audience.

- **Sentiment Analysis**: Use NLP to gauge public sentiment towards the brand or products.

- **Trend Analysis**: Identify trending topics and hashtags relevant to the brand.

- **Competitor Analysis**: Monitor competitors' social media strategies.

- **Target Audience Analysis**: Understand the demographics and interests of the followers.

- **ROI Measurement**: Track the return on investment of social media campaigns to allocate resources effectively.

57. Propose a method to forecast demand for a ride-sharing app.

For forecasting demand in a ride-sharing app:

- **Historical Data Analysis**: Analyze past ride data for patterns and trends.

- **Time Series Modeling**: Use models like ARIMA to predict future demand based on historical trends.

- **External Factors**: Incorporate external factors like weather, events, or traffic conditions.

- **Machine Learning**: Implement machine learning models to forecast demand in different areas and times.

- **Real-Time Data Incorporation**: Use real-time data for dynamic adjustments in predictions.

- **Continuous Update**: Regularly update models with new data for accuracy.

58. How would you analyze and optimize the user experience of a mobile app?

To analyze and optimize user experience in a mobile app:

- **User Behavior Tracking**: Track user interactions within the app using analytics tools.

- **Heatmaps**: Use heatmaps to understand which parts of the app are getting more attention.

- **A/B Testing**: Test different features and designs to see what works best.

- **User Feedback**: Collect and analyze user feedback through surveys or feedback forms.

- **Performance Metrics**: Monitor app performance metrics like load times and crash rates.

- **Personalization**: Implement personalized features based on user data to enhance the user experience.

59. Design a strategy to increase user engagement on a streaming platform.

To increase engagement on a streaming platform:

- **Content Recommendation**: Implement a sophisticated recommendation engine to suggest relevant content.

- **User Segmentation**: Segment users based on viewing behavior and preferences for targeted content.

- **Interactive Features**: Introduce interactive elements like polls, quizzes, or social sharing options.

- **Push Notifications**: Use personalized notifications to draw users back to the platform.

- **Exclusive Content**: Offer exclusive or early-access content to engage users.

- **Analytics-Driven Content Strategy**: Use data analytics to understand popular genres or shows and plan content accordingly.

60. How would you use data to optimize a supply chain?

To optimize a supply chain using data:

- **Demand Forecasting**: Use predictive analytics to accurately forecast product demand.

- **Inventory Optimization**: Analyze inventory levels to reduce overstock and out-of-stocks.

- **Supplier Performance**: Monitor and evaluate supplier performance using data metrics.

- **Logistics Optimization**: Use data to find the most efficient shipping routes and methods.

- **Real-Time Tracking**: Implement real-time tracking for inventory and shipments for better coordination.

- **Risk Management**: Analyze data to identify potential supply chain risks and develop mitigation strategies.

Behavioral and Situational

 61. Describe a time when you had to work with a difficult team member.

In a previous project, I collaborated with a team member who frequently missed deadlines and was resistant to feedback. To address this, I arranged a one-on-one meeting to openly discuss our project goals and how our collaboration was essential. I focused on understanding their perspective and challenges. We agreed on a more structured approach with clear, interim milestones and regular check-ins. This not only improved their engagement and reliability but also strengthened our professional relationship.

62. How do you prioritize tasks in a fast-paced environment?

In a fast-paced environment, I prioritize tasks based on urgency and impact. I use the Eisenhower Matrix to categorize tasks into urgent and important, important but not urgent, urgent but not important, and neither urgent nor important. This method allows me to focus on what needs immediate attention while also planning for tasks that contribute to long-term goals. Additionally, I maintain flexibility to adapt to changing priorities and unexpected tasks.

63. Talk about a project where you had to learn a new technology quickly.

In a recent project, I had to learn React Native for developing a cross-platform mobile application. With a tight deadline, I immersed myself in intensive learning, utilizing online tutorials, documentation, and forums. I also sought advice from colleagues experienced in React Native. By dedicating extra hours and leveraging available resources, I was able to quickly get up to speed and contribute effectively to the project, which was successfully delivered on time.

64. Describe a time when you failed and how you handled it.

In one of my early data analysis projects, I misinterpreted the data which led to incorrect conclusions. Once I realized the mistake, I immediately informed my team and supervisor, took responsibility, and revisited the analysis. I learned the importance of double-checking and validating results. This experience taught me valuable lessons in humility, integrity, and the need for meticulousness in data handling.

65. How do you stay updated with new data science trends?

To stay updated with new data science trends, I regularly:

- Read relevant online publications and blogs.
- Follow industry leaders and influencers on social media and professional networks.
- Participate in webinars and online courses.
- Attend conferences and workshops.
- Engage with the data science community through forums and local meetups.

66. Explain a complex data science concept to someone without a technical background.

Machine learning is like teaching a child through examples. Just as a child learns to identify a cat by seeing many pictures of cats, a machine learning model learns from data. It observes many examples (data), learns patterns (training), and then uses this learning to make decisions or predictions about new, unseen examples.

67. Describe a situation where you had to make a decision without all the necessary information.

In a previous role, I had to decide on a marketing strategy with incomplete customer data. I relied on available data, supplemented by industry benchmarks and competitor analysis. I also implemented A/B testing to gather more data and adjust the strategy accordingly. This approach allowed for an informed decision-making process despite the initial information gap.

68. How do you handle tight deadlines in a project?

When facing tight deadlines, I prioritize and break down tasks into smaller, manageable parts. I set clear goals for each day and maintain open communication with the team to ensure we're all aligned and on track. I also stay flexible to adjust plans as needed and ensure we're focusing on the most critical aspects of the project.

69. Talk about a successful project you led and what made it successful.

I led a project to develop a predictive analytics tool for a retail client. The project's success was attributed to a clear vision, effective team collaboration, and continuous stakeholder engagement. Regular progress updates and feedback loops ensured we were meeting the client's needs. The tool significantly improved the client's inventory management, contributing to a 20% increase in sales efficiency.

70. Describe an instance where you had to go above and beyond to solve a problem.

In a previous role, a critical system outage occurred just before a major deadline. Recognizing the urgency, I coordinated with different departments, worked extended hours, and even developed a temporary workaround to ensure minimal disruption. My proactive approach and commitment to resolving the issue not only helped meet the deadline but also resulted in developing a more robust system to prevent future outages.

7

Advanced Technical Questions

71. How do you implement a gradient descent algorithm in machine learning?

Gradient descent is an optimization algorithm used to minimize the cost function in a machine learning algorithm. It involves the following steps:

- **Initialize parameters**: Start with initial values for the parameters (weights) of the model.

- **Calculate the gradient**: Compute the gradient of the cost function with respect to each parameter.

- **Update the parameters**: Adjust the parameters in the direction that reduces the cost function. This is done by subtracting the product of the gradient and the learning rate from the current parameter values.

- **Repeat**: Continue the process iteratively until the cost function converges to a minimum.

72. Explain the concept of regularization in machine learning.

Regularization is a technique used to prevent overfitting by adding a penalty to the loss function. The most common types are L1 (lasso) and L2 (ridge) regularization. L1 penalizes the absolute value of weights, leading to sparsity and feature selection. L2 penalizes the square of the weights, which tends to distribute the error among all terms. Regularization helps in enhancing the generalization capabilities of the model.

73. How would you implement a recommendation system for a large dataset?

For a large dataset, a recommendation system can be implemented using:

- **Collaborative Filtering**: Leveraging user-item interactions and finding similarities between users or items.

- **Matrix Factorization Techniques**: Such as Singular Value Decomposition (SVD), which are efficient for large datasets.

- **Deep Learning**: Using neural networks to capture complex patterns in large-scale data.

- **Scalable Algorithms**: Employing algorithms and data structures that handle large volumes of data efficiently.

- **Distributed Computing**: Using platforms like Apache Spark to parallelize computations.

74. What are convolutional neural networks and where are they used?

Convolutional Neural Networks (CNNs) are a type of deep learning model primarily used for image processing and analysis. They consist of convolutional layers that apply filters to raw pixel data to extract and learn features from images. CNNs are used in applications like image classification, object detection, face recognition, and more.

75. Explain time series analysis and its applications.

Time series analysis involves analyzing time-ordered data points to extract meaningful statistics and characteristics. It's used to forecast future values based on previously observed values. Applications include stock market analysis, economic forecasting, weather prediction, and any domain where data is collected over time intervals.

76. Discuss the advantages and disadvantages of different loss functions.

- **Mean Squared Error (MSE)**: Good for regression, sensitive to outliers.

- **Mean Absolute Error (MAE)**: Less sensitive to outliers compared to MSE, used in regression.

- **Cross-Entropy**: Used for classification, penalizes incorrect classifications heavily.

- **Hinge Loss**: Used for support vector machines, not as sensitive to outliers as cross-entropy.

77. How do you handle imbalanced datasets in classification problems?

To handle imbalanced datasets:

- Use resampling techniques: Oversample the minority class or undersample the majority class.

- Apply synthetic data generation methods like SMOTE.

- Use appropriate evaluation metrics like F1-score, Precision-Recall AUC instead of accuracy.

- Employ algorithmic approaches like cost-sensitive learning or anomaly detection methods.

78. What are the challenges in working with large-scale data?

Challenges include:

- **Storage and Processing**: Requires substantial storage and computing power.

- **Performance**: Algorithms may not scale efficiently with data size.

- **Data Quality**: Managing noise, outliers, and missing values becomes more complex.

- **Analysis Complexity**: More data can lead to more complex models, which can be harder to interpret.

79. Explain the concept of dimensionality reduction and its importance.

Dimensionality reduction involves reducing the number of input variables in a dataset. Techniques like Principal Component Analysis (PCA) or t-Distributed Stochastic Neighbor Embedding (t-SNE) are used. It's important because it helps in dealing with the curse of dimensionality, reduces computational cost, and can improve model performance by removing irrelevant features.

80. How would you approach building a model to detect anomalies in time-series data?

To detect anomalies in time-series data:

- Use statistical models like ARIMA for modeling normal behavior and flag deviations.

- Implement machine learning models like Isolation Forest or Autoencoders.

- Leverage deep learning for complex patterns, e.g., LSTM networks.

- Apply moving averages or rolling windows to identify unusual points.

- Employ domain-specific thresholds to determine anomalies.

Data Analysis and Visualization

81. How do you ensure the quality of your data analysis?

To ensure quality in data analysis:

- **Data Cleaning**: Address missing values, outliers, and inconsistencies.

- **Data Validation**: Cross-check data sources and ensure data integrity.

- **Exploratory Data Analysis (EDA)**: Understand data distributions and relationships.

- **Robust Statistical Methods**: Use appropriate and validated statistical methods.

- **Peer Review**: Have analyses reviewed by colleagues for errors or biases.

- **Reproducibility**: Document processes to allow others to replicate the analysis.

82. Explain different ways to visualize time-series data.

Time-series data can be visualized using:

- **Line Charts**: To display trends over time.

- **Area Charts**: Similar to line charts but with the area below the line filled in, emphasizing the magnitude.
- **Bar Charts**: Useful for comparing time-based data categories.
- **Scatter Plots**: For examining relationships between different variables over time.
- **Heatmaps**: Useful for visualizing complex data patterns and variations across multiple time periods.

83. Describe how you would use data visualization to present insights to non-technical stakeholders.

For non-technical stakeholders:

- Use simple and clear visualizations like bar charts, line graphs, and pie charts.
- Avoid technical jargon and explain the context and relevance of the data.
- Use annotations, labels, and legends for clarity.
- Highlight key insights and trends directly on the visuals.
- Use interactive dashboards for stakeholders to explore the data themselves.

84. What are the key considerations when choosing a visualization type?

When choosing a visualization type, consider:

- **Data Type and Scale**: Different visuals suit different data types (categorical, numerical) and scales.
- **Objective**: What you want to communicate (trends, distributions, relationships).
- **Clarity and Simplicity**: The visualization should be easy to understand.

- **Audience**: Tailor the visualization to the knowledge and interests of your audience.

85. How do you handle outliers in your data?

Handling outliers involves:

- **Detection**: Identify outliers using statistical methods like IQR or Z-scores, or visualization tools.

- **Assessment**: Determine if they represent errors or genuine extreme values.

- **Treatment**: Depending on the assessment, outliers can be removed, corrected, or left as is. In some cases, transformations or robust methods are used.

86. Explain the process of hypothesis testing in a data analysis project.

Hypothesis testing involves:

- **Formulating Hypotheses**: Define a null hypothesis (H0) and an alternative hypothesis (H1).

- **Selecting a Test**: Choose an appropriate statistical test based on data type and distribution.

- **Setting Significance Level**: Commonly set at 0.05.

- **Calculating Test Statistic**: Based on the chosen test.

- **Decision Making**: Compare the test statistic with critical values to accept or reject H0.

87. What tools do you use for data visualization and why?

Common data visualization tools include:

- **Tableau**: For interactive and powerful data visualizations.

- **Microsoft Power BI**: Integration with Microsoft products and ease of use.

- **Python Libraries (Matplotlib, Seaborn, Plotly)**: For flexibility and customization in data analysis workflows.

- **R (ggplot2)**: For extensive data visualization capabilities in statistical analysis.

88. Describe a scenario where data visualization provided critical insights into a problem.

In a retail business, data visualization using heatmaps and line charts of sales data revealed seasonal trends and customer preferences. This insight helped in optimizing inventory and marketing strategies, leading to increased sales and customer satisfaction.

89. How do you validate the results of your data analysis?

To validate results:

- **Cross-Validation**: Use techniques like k-fold cross-validation.

- **Out-of-Sample Testing**: Test predictions on a separate dataset.

- **Statistical Significance Testing**: Ensure results are statistically significant.

- **Sensitivity Analysis**: Check how results vary with changes in assumptions or inputs.

- **Peer Review**: Have others review the analysis.

90. Discuss a technique for effectively communicating complex data analysis to a lay audience.

To communicate complex data analysis:

- **Simplify Concepts**: Break down complex ideas into simpler terms.

- **Use Analogies**: Relate concepts to everyday experiences.

- **Visual Aids**: Employ clear and straightforward visualizations.

- **Storytelling**: Present data in a narrative form to engage the audience.

- **Focus on Key Insights**: Highlight the most important findings and their implications.

Technical Theory and Concepts

91. Explain the concept of Big O notation.

Big O notation is a mathematical notation used in computer science to describe the performance or complexity of an algorithm. Specifically, it characterizes the time complexity and space complexity of an algorithm in terms of the worst-case or upper limit as the input size grows. For example, an algorithm with a time complexity of O(n) means that the time it takes to complete the task grows linearly with the input size.

92. Discuss the differences between SQL and NoSQL databases.

- **SQL Databases**: These are relational databases that use structured query language (SQL) for defining and manipulating data. They are characterized by a fixed schema and are best suited for complex queries. Examples include MySQL, PostgreSQL, and SQL Server.

- **NoSQL Databases**: These are non-relational or distributed databases known for their flexibility to handle large volumes of unstructured data. They have dynamic schemas for unstructured data and are suitable for horizontal scaling. Types include document (MongoDB), key-value (Redis), wide-column (Cassandra), and graph databases (Neo4j).

93. What are the main components of a distributed computing system?

A distributed computing system typically includes:

- **Nodes**: Individual computers or servers within the network.

- **Network Communication**: The method for information exchange between nodes (like HTTP, TCP/IP).

- **Distributed File System**: Allows data to be stored in an easily accessible format across a network of machines.

- **Task Scheduler**: Allocates tasks to different nodes.

- **Fault Tolerance Mechanisms**: Ensure the system continues to operate in case of node failures.

94. Explain how a blockchain works and its potential applications.

Blockchain is a decentralized, distributed ledger technology that records transactions across many computers securely and in a verifiable and permanent way. Once a record is added to the chain, it is very difficult to change. Each block contains a cryptographic hash of the previous block, a timestamp, and transaction data. Applications include cryptocurrency (like Bitcoin), smart contracts, supply chain management, voting systems, and more.

95. What are the challenges of working with real-time data?

Challenges of working with real-time data include:

- **Volume and Velocity**: Managing high volumes of data at high speed.

- **Data Quality and Accuracy**: Ensuring the data is accurate and reliable.

- **Latency**: Minimizing delays in data processing and analysis.

- **Scalability**: Ability to handle increasing loads of data.

- **Data Integration**: Combining data from various sources.

96. Describe the MapReduce programming model.

MapReduce is a programming model for processing and generating large datasets with a parallel, distributed algorithm on a cluster. It consists of two steps:

- **Map Step**: Processes and converts input data into key-value pairs.

- **Reduce Step**: Aggregates and processes the key-value pairs to output the final result. MapReduce allows for scalable and efficient processing of vast amounts of data.

97. What is the significance of data normalization?

Data normalization is a process in databases to structure data to reduce redundancy and improve data integrity. It involves organizing fields and tables of a database to minimize dependency and duplication. Normalization helps streamline the database design, making it more efficient and easier to maintain and update.

98. Discuss the concept of data warehousing.

Data warehousing is the electronic storage of a large amount of information by a business, in a manner that is secure, reliable, easy to retrieve, and easy to manage. It's designed to facilitate reporting and analysis. Key components include:

- **Central Repository**: Aggregates data from multiple sources.

- **Data Integration**: Involves ETL (Extract, Transform, Load) processes.

- **Historical Data Storage**: Maintains historical data for trend analysis.

- **Business Intelligence Tools**: Used for querying and analyzing data.

99. Explain the CAP theorem in distributed systems.

The CAP theorem, also known as Brewer's theorem, states that a distributed database system can only simultaneously provide two of the following three guarantees:

- **Consistency**: Every read receives the most recent write or an error.

- **Availability**: Every request receives a response, without guaranteeing it contains the most recent write.

- **Partition Tolerance**: The system continues to operate despite arbitrary partition failures. The theorem is a fundamental principle to understand the limitations and trade-offs in distributed systems.

100. What are the key principles of data ethics and privacy?

Key principles include:

- **Consent**: Obtaining permission from individuals before collecting their data.

- **Anonymity and Privacy**: Protecting personal identity and sensitive information.

- **Transparency**: Being open about how data is collected, used, and shared.

- **Security**: Ensuring data is securely stored and protected from breaches.

- **Fairness and Non-discrimination**: Avoiding biases in data collection and analysis.

- **Accountability**: Taking responsibility for the ethical use of data.

10

Interview-Specific Questions

101. How would you prepare for a technical interview at a FAANG company?

To prepare for a technical interview at a FAANG company:

- **Study the Basics**: Review fundamental concepts in your field (e.g., algorithms, data structures, system design).

- **Practice Coding**: Regularly solve coding problems on platforms like LeetCode or HackerRank.

- **Understand the Company's Technology Stack**: Familiarize yourself with the technologies and tools the company uses.

- **Mock Interviews**: Engage in mock interviews to simulate the interview environment.

- **Review Past Projects**: Be prepared to discuss your past work and projects in detail.

- **Learn About the Company Culture**: Understand the company's core values and work culture.

102. What are the key qualities FAANG companies look for in a data scientist?

FAANG companies typically look for:

- **Technical Expertise**: Strong skills in programming, statistics, machine learning, and data analysis.
- **Problem-Solving Ability**: The capacity to approach complex problems methodically and creatively.
- **Communication Skills**: Ability to explain technical concepts to non-technical stakeholders.
- **Teamwork and Collaboration**: Collaborating effectively with team members.
- **Adaptability and Continuous Learning**: Staying updated with industry trends and adapting to new technologies.
- **Impactful Work**: Demonstrating how past work has had a significant impact.

103. How do you approach solving unfamiliar problems in an interview setting?

When faced with unfamiliar problems:

- **Stay Calm**: Keep a calm and focused mindset.
- **Break Down the Problem**: Divide the problem into smaller, manageable parts.
- **Think Aloud**: Explain your thought process as you work through the problem.
- **Draw from Past Experience**: Relate the problem to anything similar you've encountered before.
- **Ask Clarifying Questions**: Ensure you fully understand the problem and ask questions if necessary.

- **Propose a Methodical Solution**: Even if you're not sure, propose a logical approach.

104. What strategies do you use to manage stress in interviews?

To manage stress:

- **Prepare Thoroughly**: Being well-prepared boosts confidence.

- **Practice Relaxation Techniques**: Like deep breathing or meditation before the interview.

- **Visualize Success**: Imagine a positive outcome to build confidence.

- **Stay Healthy**: Get enough sleep, eat healthily, and exercise.

- **Positive Self-talk**: Remind yourself of your capabilities and past successes.

- **Arrive Early**: Give yourself plenty of time to relax and get comfortable with the setting.

105. Discuss how to effectively communicate your thought process during an interview.

To communicate your thought process:

- **Speak Clearly and Concisely**: Be articulate in explaining your thoughts.

- **Logical Flow**: Present your ideas in a structured, logical manner.

- **Engage with the Interviewer**: Ask if they follow your reasoning and be open to feedback or hints.

- **Use Examples**: Illustrate your points with relevant examples.

- **Show Enthusiasm**: Let your interest and passion for the subject show.

106. What makes you a good fit for a data science role in a FAANG company?

A good fit for a data science role in a FAANG company may be someone with:

- **Strong Technical Skills**: In areas like programming, statistics, and machine learning.

- **Problem-solving Ability**: Proven skills in tackling complex, real-world problems.

- **Adaptability**: Ability to learn and apply new technologies and methodologies.

- **Collaborative Mindset**: Experience in working effectively in diverse teams.

- **Innovative Thinking**: Bringing new ideas and perspectives to the table.

- **Alignment with Company Values**: Sharing the company's vision and values.

107. How do you stay motivated during the job search and interview process?

To stay motivated:

- **Set Goals**: Have clear, achievable objectives.

- **Regular Breaks**: Avoid burnout by taking regular breaks.

- **Seek Feedback**: Learn from each interview experience, whether successful or not.

- **Stay Positive**: Maintain a positive attitude and perspective.

- **Network**: Engage with professionals in the field for support and advice.

- **Celebrate Small Wins**: Acknowledge and celebrate progress and achievements.

108. What are the most common mistakes candidates make in data science interviews?

Common mistakes include:

- **Lack of Preparation**: Not being well-prepared for technical questions.

- **Not Understanding the Question**: Failing to ask clarifying questions when needed.

- **Poor Communication**: Not clearly articulating thought processes.

- **Neglecting Soft Skills**: Focusing only on technical skills and not demonstrating teamwork or communication skills.

- **Not Showcasing Impact**: Failing to explain how their work has had a tangible impact.

109. How do you demonstrate your teamwork and collaboration skills in an interview?

To demonstrate teamwork and collaboration:

- **Share Specific Examples**: Talk about times you've successfully worked in a team.

- **Highlight Communication Skills**: Show how you communicate and collaborate with team members.

- **Discuss Conflict Resolution**: Share how you've handled disagreements or challenges in a team.

- **Emphasize Team Achievements**: Focus on the team's success, not just individual contributions.

- **Active Listening**: Demonstrate good listening skills during the interview itself.

110. What questions should you ask the interviewer to understand the role better?

Questions to ask:

- **Role-Specific Questions**: What are the day-to-day responsibilities of this role?

- **Team Structure**: Can you tell me more about the team I'll be working with?

- **Success Measurement**: How is success measured for this position?

- **Growth Opportunities**: What opportunities for professional growth does the company offer?

- **Company Culture**: Can you describe the company culture and values?

- **Future Projects**: Are there any upcoming projects or challenges the team is facing?

Conclusion

As we draw the curtains on "Ace the Data Science Interview: Unveil the Secrets of 100 Real Questions from the World's Leading Tech Giants," it is important to reflect on the journey we have embarked upon together. Through the pages of this book, we ventured into the intricate and challenging world of data science interviews at the most esteemed tech giants. We navigated through a myriad of questions, from the fundamental to the complex, each designed to not only prepare you technically but also to sharpen your analytical and problem-solving skills.

The road to a successful career in data science, especially within the walls of FAANG companies, is undeniably rigorous. It demands more than just technical knowledge; it requires resilience, adaptability, and a continuous thirst for learning. This book has aimed to equip you with the tools and confidence to face these challenges head-on.

Remember, every question explored and every concept understood brings you a step closer to your goal. The real-world interview questions discussed in this book are your arsenal in this competitive field. They are a window into what top companies seek in their ideal candidates and a blueprint for what you should aspire to be.

As you close this book and proceed towards your interviews, carry with you the lessons learned, the strategies developed, and the knowledge gained. Let them guide you in presenting your best self to your future employers.

We wish you the very best as you step forward in this exciting phase of your career. May your journey be rewarding, your learning continuous, and your successes many. Good luck as you endeavor to not just ace your data science interviews, but to carve a niche for yourself in the world of technology and innovation.

With all the insights and preparations, you are now ready to turn challenges into opportunities. Go forth and conquer!

Best wishes and good luck,

AI Publishing

From the Same Publisher

Python Machine Learning
https://bit.ly/3gcb2iG

Python Deep Learning
https://bit.ly/3gci9Ys

Python Data Visualization
https://bit.ly/3wXqDJl

Python for Data Analysis
https://bit.ly/3wPYEM2

Python Data Preprocessing

https://bit.ly/3fLV3ci

Python for NLP

https://bit.ly/3chlTgm

10 ML Projects Explained from Scratch

https://bit.ly/34KFsDk

Python Scikit-Learn for Beginners

https://bit.ly/3fPbtRf

Data Science
with Python

https://bit.ly/3wVQ5iN

Statistics
with Python

https://bit.ly/3z27KHt